All MEYER LEMONS are my Children

Dr. Chester N. Roistacher

Copyright © 2018 Dr. Chester N. Roistacher

All rights reserved.

ISBN 10: 19850996
ISBN-13: 978-1985090996

DEDICATION

In the late 1960's while working on thermotherapy of citrus for the elimination of viruses I came across an abstract of a paper by Professor Lin Kung-Hsiang, a Chinese Plant Pathologist, and had his abstract translated to English. The abstract reported on a creative new method for thermotherapy using moist-hot air on plants in a special chamber. This was a new technique to me instead of the harsher hot water treatments which proved too severe and were mostly unsuccessful. Subsequently I devised equipment duplicating his method for elimination viruses in citrus and by using his new method was able to eliminate the tatter leaf virus from the Meyer lemon. This tatter leaf virus was very destruction to citrus worldwide and especially in China. As a result of using his technique making available a virus-free Meyer lemon, in 1975 a law was passed in the State of California decreeing that all Meyer lemons must be produced from virus-free budwood which had been developed by the moist-hot air technique.

In 1982 I was invited to China to lecture at the Central China Agricultural University in Wuhan. Prior to my lectures, I had the privilege of touring the country. Having learned that Prof. Lin was in Canton (Quangdong) and while visiting that city I asked if I could see him. They brought him to my hotel room and all evening long we enjoyed pleasant discussions on plant pathology and reminisced on our college days since we were both graduates of Cornell University. I took a picture of him seated in my hotel room (shown below). At that time he was under treatment for cancer. He was very pleased when I told him how indebted I was to him for his original research on thermotherapy using the moist-hot air technique. The next morning a letter was delivered to me from Dr. Lin part of which reads as follows:

"If you should have a chance to mention my work and make your evaluation of the heat treatment method I developed as you did last night, and if you consider it appropriate, it will facilitate my work greatly. We have a Chinese saying, "Local ginger does not taste hot".

Prof. Lin Hung-hsiang in 1982   A statue of Prof. Lin Hung-hsiang in front of the Citrus Research Institute in Beibei, China

Subsequently, when lecturing in Wuhan, I gave him due credit for the elimination of the tatter leaf virus from citrus by using his hot–moist air treatment. Dr. Lin died of cancer in 1985. In 1991 I sent the picture above to his wife Lucille Lin who was then living in Syracuse, New York. In response, Mrs. Lin wrote and thanked me and mentioned the difficulties they endured during the Cultural Revolution in China and wrote of my role in liberating him and his family.

One never knows the paths of the heart and how we touch one another in ways that are strange and sometimes rewarding. The rehabilitation of Professor Lin was apparent when, in the introductory foreword to the Proceedings of the Asia Pacific International Conference on Citriculture at Chiang Mai, Thailand in 1990, Dr. Lin was honored by having the Proceedings dedicated to his memory. In addition, the statue of Prof. Lin shown above was erected in front of the Citrus Research Institute, Chinese Academy of Agricultural Sciences/Southwest University, Beibei, Chongqing.

Chester Roistacher

CONTENTS

Acknowledgments	i
The Meyer Lemon, My Story	1-14
Letter to Mrs. Lucille Lin	15
Letter from Mrs. Lucille Lin	16
Letter from Dr. Lin Kung-Hsiang	19
Obituary Dr. Lin Kung-Hsiang	22
Books to be published by Dr. C. N. Roistacher	25

The events you are about to read are based on a true story.

In the mid 1970's after appropriate hearings, and then by law, all Meyer lemon trees produced in California had to be certified budwood properly tagged as the "Certified Improved Meyer lemon."

This book is dedicated to the memory of Professor Lin Kung-Hsiang whose pioneering work of the use of moist-hot air enabled the eradication of the viruses in the Meyer lemon and thus leading to the development of the virus free Meyer lemon.

The Meyer Lemon

In the early spring of 2015 while walking in my backyard admiring my garden, feeding birds and admiring my citrus trees, I noticed the fruit growing on my Meyer lemon tree was glowing with its winter coat of yellow. I stopped, and as I looked this tree over I realized there is a story about this tree which should be told.

A thought occurred to me that for the past forty plus years, all Meyer lemon trees grown in California, and probably throughout the world are my children. The Meyer lemon came from a single budstick that I had treated with moist-hot air to rid it of viruses which had plagued the trees. My method of treatment resulted in a virus-free Meyer lemon which was named, "The Certified Improved Meyer Lemon."

In my book, "My Life in Agriculture," I share the story of the research leading to how we gave a fever to a Meyer lemon budstick similar to when your body temperature increases to rid itself of a virus or a bacteria. In this same way the viruses in the Meyer lemon budstick were eliminated. Of interest, the technology for this thermo-therapy came from a Chinese scientist, Professor Lin Kung-Hsiang, and by my honoring him during a lecture given in Wuhan, China in 1982 and by giving him due credit for his pioneering research on thermotherapy, I had liberated him and his family from the difficult suppressions during the days of the 'Cultural Revolution' in China where he had suffered much.

This is my story of the Meyer lemon.

How it all began.

After graduating Cornell University in 1949, I began working as a technician at UCLA for Dr. Kenneth Baker, the world's foremost Plant Pathologist on the diseases of Floricultural crops.

My first job was working with the gladiolus; a flower being grown in an area around Vista, California. These gladiolus flowers were for commercial production and were grown outdoors. However, their bulbs were subject to a serious fungus disease called *Fusarium*. This disease ravished the gladiolus industry. A field planted for the gladiolus flower would last only a year or two, and then abandoned and new plantings were done in a new region in non-infected soil with the belief that the gladiolus flower would not be infected in the new soil.

Plantings were continually moved from one area to another in the Vista area since replanting gladiolus bulbs in the same area would infect the new plantings. There was no known way to prevent this infection. I was tasked with finding a treatment for the fungus ravishing this lovely flower.

Left: Showing a gladiolus bulb and little cormels. These were the cormels which I subjected to hot water treatment.

BEGINNING HEAT THERAPY FOR GLADIOLUS .

My job was to try to rid the bulbs or cormels of this *Fusarium* fungus. Experiments were begun by treating the small infected cormels in hot water. As shown above, cormels are the small pea sized offshoots from the bulbs and were ideal for hot water treatment since they were found more tolerant to heat than the large bulbs. I found that immersing the cormels in hot water was too drastic a treatment which killed them. I then began experimenting using various means to pre-condition the cormels by subjecting them first to warm temperatures. Following reports in the literature, I found that by adding 5% alcohol to the hot water bath was effective in killing *Fusarium* and not killing the bulbs.

By the use of various pre-conditioning methods and with the addition 5% alcohol, success was achieved in eliminating the *Fusarium* fungus from the cormels. This was published in the 'Journal of Hilgardia' as shown.

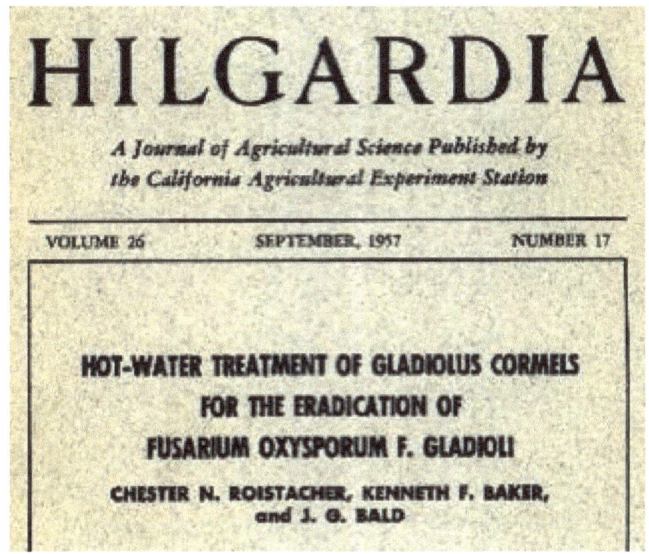

Publication in the scientific Journal Hilgardia.

I began working at the University of California at Riverside (UCR) in the 1960's with citrus.

In the 1960's I began working intensely on a project to rid citrus of its various pathogens. At that time the only known way to bypass viruses and other pathogens in citrus was by growing citrus from seed. There are two kinds of trees which are produced from seed: the Gametic and the Nucellar. Seedlings grown from the Gametic nucleus contain genes of both parents resulting in trees and fruit which were different from the original. The Nucellar layer of tissue which surrounds the embryo is a group of cells which can germinate independently of the nucleus and produce a tree which would be pathogen-free.

However, the resulting trees had many serious problems. For example, they took a long time to outgrow their juvenility, they would come into production later and the quality of fruit might be less than the original. Nucellar trees were thorny with upright growth, excessive tree vigor, with a tendency towards alternate bearing and an unequal distribution of fruit on the tree. In view of these problems a new way was needed to rid the pathogens from citrus plants other than going through the nucellar.

Shown is a diagram explaining the nucellar. Progagations from budwood originating from the gametic tree (below) will pass pathogens in its budwood to its progeny trees. However, as shown in the tree above, most citrus pathogens (viruses, viroids or bacteria) will NOT pass through the nucellus tissue inside its seed. Therefore, citrus trees developed from nucellar seedlings are usually virus-free. However, as mentioned, there are problems associated with trees produced from nucellar.

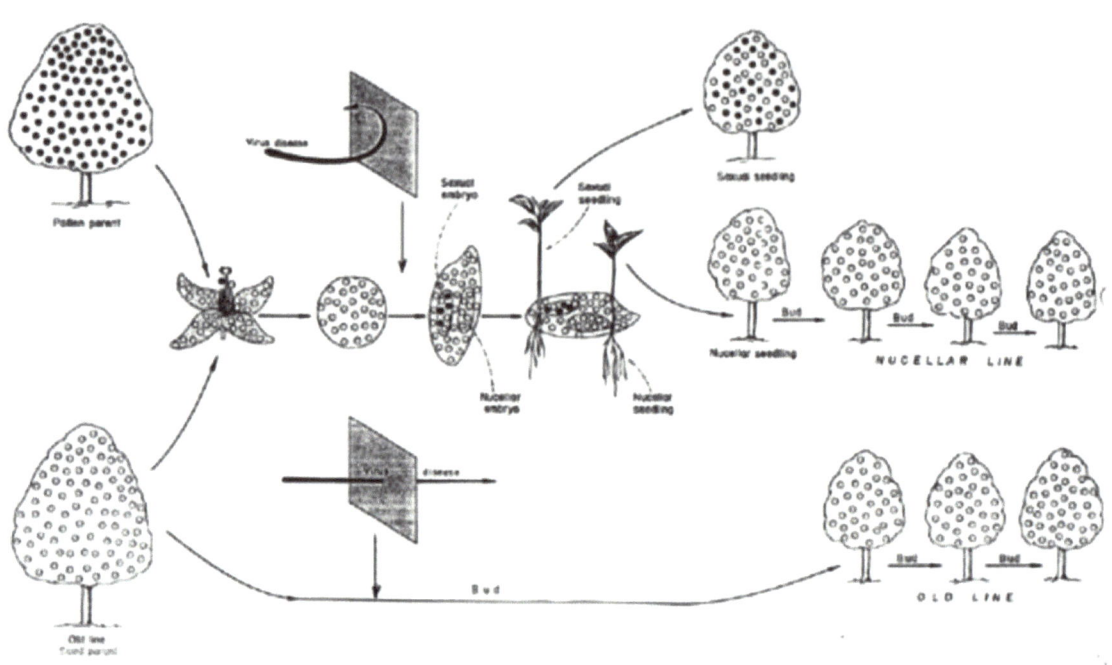

NUCELLAR LINE

PROBLEMS WITH THE NUCELLAR
- LONG TIME TO OUTGROW JUVINILITY
- LONG TIME TO COME INTO PRODUCTION
- POORER QUALITY OF NUCELLAR FRUIT
- EXCESSIVE THORNINESS
- UPRIGHT TREE GROWTH HABIT
- EXCESSIVE TREE VIGOR
- ALTERNATE BEARING TENDENCY
- UNEQUAL FRUIT DISTRIBUTION
- DETERIORATION OF FRUIT QUALITY

Beginning heat therapy of citrus budwood by treating budwood in hot water.

In the 1950's scientists in our field of citrus virology published on the use of hot water as therapy for eliminating viruses in citrus budwood. In nearly all reports, hot water treatment was found to be too drastic and the budwood would be cooked before the internal pathogens were killed. Going back to my early work in the 1950's where I had some success by pre-conditioning gladiolus cormels, I began studies on pre-conditioning citrus budwood so it could withstand and survive the hot water treatment. I found that by subjecting citrus plants to very warm temperatures in the hot part of our greenhouse before subjecting their budwood to hot water I had a successful outcome and achieved eliminating a few viruses, and published on this (Roistacher and Calavan, 1972). However, I found that hot water was still a too drastic a treatment for treating citrus budwood and many viruses could not be eliminated.

Thermotherapy for the eradication of pathogens from seed and plant tissue has been used for centuries and is still an ecological and effective means of eradicating pathogens from propagative budwood in many crops, especially sugar cane.

The early attempts using hot water to eradicate citrus pathogens mostly failed (Fawcett and Cochran, 1941; Grant et al., 1950; Price and Knorr, 1956). Grant (1957), successfully eliminated the Citrus tristeza the Citrus Psorosis viruses using hot air in chambers at 35-42°C for 78-107 days.

Roistacher and Calavan (1972) showed that unless budwood was pre-conditioned to heat, it would mostly fail to survive hot water treatment at 48 and 50°C.

TABLE 1. THE EFFECT OF PRECONDITIONING ON SURVIVAL OF PINEAPPLE SWEET ORANGE BUDWOOD TREATED IN HOT WATER

Budwood source	Hot water treatments Temp. (°C)	Time (hrs)	Bud survival (of 20)	Bud growth
Preconditioned[a]	Control		20	17/20
	48	2	20	18/20
	48	3	18	15/18
	50	2	20	15/20
	50	3	16	6/16
Nonpreconditioned[b]	Control		19	15/19
	48	2	15	7/15
	48	3	9	7/9
	50	2	2	1/2
	50	3	0	

a. Budwood was preconditioned for 28 weeks at temperatures of 30-31° minimum and 38-42°C maximum.
b. Normal temperatures were 20-22° minimum and 27-33°C maximum.

Table 1, from Roistacher and Calavan (1972) showing general failure of non-preconditioned budwood to hot water.

Moist-hot air for thermotherapy - A discovery by Professor Lin Kung-Hsiang.

My supervisor Dr. E.C. Calavan alerted me to a reference transcribed from the Chinese on the work of Professor Lin Kung-Hsiang where he used moist-hot air for the treatment of citrus budwood and was successful in eliminating the yellow shoot disease caused by the greening disease bacteria (huanglongbing -see Appendix 1 Page 21). His use of moist-hot air was new and revolutionary. A comparison could be made by taking a steam sauna bath where our bodies are able to tolerate temperatures of 160° to 180°F (71 to 82°C). However, taking a bath in hot water at these temperatures would certainly scald us. Experiments were then initiated using Professor Lin's moist-hot air technique for treating citrus budwood.

Figure 1: Above left - A plastic cage which was placed in the hot room of our greenhouse. Citrus plants inside this cage were preconditioned by the heat inside the cage. Above right – showing the old incubator with a hot plate as a source of heat. The temperature was controlled by a thermostat and the air circulated by a fan.
Below – Showing the moist-hot air treatment: Citrus budsticks were placed inside a plastic container with a small amount of water on the bottom to create a moist environment. The budsticks were subjected to heat at 122 °F (50 °C) inside the plastic container and temperatures observed with a thermometer.

Citrus plants infected with pathogens were pre-conditioned by placing them inside a plastic cage (Figure 1) on a bench inside a section of our greenhouse which was kept at a warm temperature. Extra heat was applied inside this plastic cage by using a small electric heater. A small fan circulated the air inside the cage and a thermostat regulated the temperature. Temperatures inside this cage reached 100° to 104 °F (38 to 40°C) during the day. Plants were held in this warm cage for one to two months. Budwood from these pre-conditioned plants were then subjected to the moist-hot air treatment as illustrated in Figure 1.

Treatment of Meyer Lemon budwood in moist-hot air.

In the midst of my research using moist-hot air treatment and while conducting an experiment on thermotherapy using an old incubator as illustrated in Figure 1, a citrus nurseryman from Central California was visiting our Rubidoux indexing greenhouse at UCR. He was curious about what I was doing and I explained that I was heat treating budwood in order to eliminate viruses and I showed him the incubator which was then in operation. He asked if this technique might be used to rid the Meyer lemon of its viruses, one of which was the serious Citrus tristeza virus and the other was the Citrus tatter leaf virus.

At that time, the Central Valley of California was under quarantine for the Citrus tristeza disease and since all Meyer lemons present in the Central Valley contained the tristeza virus, a Meyer lemon eradication program was then in effect throughout the Central Valley. At that time and by law, all Meyer lemons in the Central Valley were being eradicated. Since the Meyer lemon was almost exclusively a home backyard tree, personnel of the California Tristeza Control Agency were sent out in an intensive house to house search to locate and destroy all Meyer lemon trees.

The Meyer lemon was introduced into the United States from Beijing, China in 1908 by the plant hunter Frank Meyer. In California it is primarily grown as a home-yard tree. The tree is cold hardy and produces an abundant of colorful fruit each year. As mentioned, the introduced Meyer lemon from China was found to contain two viruses: tristeza and tatter leaf. When I had been consulting in Zhejiang Province, China in 1986 I had observed that both of these viruses were present and were widely distributed throughout the Province.

Also, all Meyer lemon trees as well as the sweet orange trees infected with the tatter leaf virus and were on trifoliate orange rootstock growing in Zhejiang Province were infected. The infected trees with the trifoliate orange rootstock showed a severe decline in the presence of the tatter leaf virus as shown in Figure 2. Similarly, propagations from the introduced Meyer lemon from China into California and worldwide were all found to be infected with both viruses and could not be successfully propagated on trifoliate or trifoliate hybrid rootstocks and would decline.

Figure 2. Showing decline of oranges on trifoliate orange rootstock due to the tatter leaf virus when I visited Zhejiang Province, China in 1986.

In response to the inquiry from the visiting citrus nurseryman, experiments were initiated to see if this new moist-hot air technique could eliminate the tatter leaf and tristeza viruses in our infected Meyer lemon[1] tree. Therefore, a plant of Meyer lemon was placed inside our plastic pre-conditioning chamber in our warm greenhouse room as shown in Figure 1. After ten weeks of pre-conditioning at warm temperatures in the warm room in our plastic cage, small budsticks were cut from the pre-conditioned Meyer lemon plant and put inside our incubator as illustrated in Figure 1.

Treatment of the Meyer lemon budwood in moist-hot air was at 122°F (50°C) inside the plastic container with a small amount of water at the bottom and monitored by a thermometer. The temperature inside the incubator was controlled by a thermostat. During the day of the moist-hot air treatment, a single budstick was removed from the plastic container at 3, 4, 5, 6, 7, and 11 hours respectively.

Since there was no evidence of the budsticks being cooked, the experiment was continued overnight and into the morning, and after 22 hours of treatment, a final budstick was removed for indexing. All of the budwood survived at all of the hours of treatment. Buds from the surviving budsticks were propagated on seedlings and the plants produced were indexed. Ten of these plants were found free of the tatter leaf virus by this moist-hot air treatment. Indexing was done by grafting tissue from the heat treated plants to citrange seedlings which were sensitive to the tatter leaf virus. If no symptoms appeared in the index plants, but symptoms appeared in the positive control plants, the tatter leaf virus was eliminated!

---

[1] *The Meyer lemon plant held at the Rubidoux screen house and used in these heat treatment experiments was obtained by Dr. Meryl I Wallace from Texas, and was free of the Citrus tristeza virus.*

The table below shows the results of this experiment from Roistacher and Calavan (1972). Ten buds had survived the moist-hot air treatment.

TABLE 3. BUD SURVIVAL AND INACTIVATION OF TATTERLEAF-CITRANGE STUNT VIRUS IN PRECONDITIONED MEYER LEMON BUDWOOD TREATED IN MOIST AIR AT 50° C

Hrs. at 50°C	Buds surviving (of 8)	Symptomless citrange indicators (of 4)
0	8	0
0	8	1
3	8	0
4	8	1
5	8	1
6	8	1
7	8	3
11	8	0
22	8	3

Return of the Meyer lemon to Central California.

In order to bring this new heat-treated and virus-free Meyer lemon back into the Central valley of California, hearings were held by the California Department of Food and Agriculture on how this should be done. It was decided that the new disease-free Meyer lemon should have a new name and it would be called "*The Certified Improved Meyer lemon*". All plants propagated and bred from this virus-free selection would be so tagged. Citrus nurserymen were then invited to an official hearing and were asked how long a period of time they would need before their stock of existing Meyer lemon plants could be sold. They agreed that after a period of two years, only the properly tagged `*Certified Improved Meyer lemon*' could legally be sold from their nurseries in California.

Figure 3: Left: The new Meyer lemon tag. Right: the new virus-free Meyer lemon tree at the Lindcove Foundation Block.

The first 1,000 trees of Meyer lemon were certified on December 11, 1975. Budwood for these trees were produced from the virus-free Meyer lemon trees in the University of California Foundation Block at the Lindcove Field Station in Central California (Figure 3). All new virus-free Meyer trees produced in California were in compliance with the Department's "Regulations for Registration and Certification of Certified Improved Meyer lemon trees."

The Meyer lemon would now be permitted to be planted back in the Central Valley of California. Thus, all Meyer lemon trees produced since 1975 were derived from the buds which I had treated by the moist-hot air in the crude incubator. As mentioned, the initial impetus for this research coming from a visiting citrus nurseryman from California's Central Valley and the technology of moist-hot air treatment derived from the work of Professor Lin Kung-Hsiang.

Propagation – Producing a Tree by Grafting.

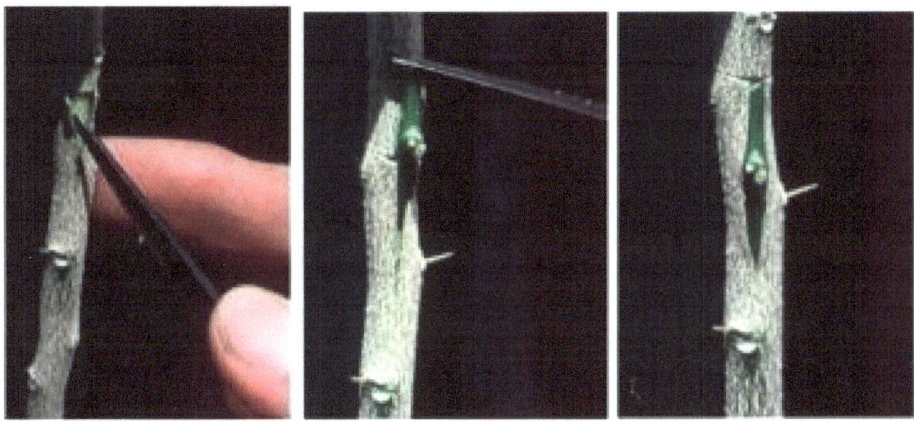

1-The T-Cut   2-Inserting the Bud   3-The Bud fully inserted

4-Wrapping the Bud & leaving the eye exposed   5-The Bud growing to form a tree

If we assume that the bud that is shown being inserted in the rootstock seedling came from a bud stick that had been treated in moist-hot air as shown in Figure 1, then the tree grown from that bud, as shown in the tree growing at the Lindcove Foundation Block in Figure 3, would be my child and is the result of the work done by Professor Lin, using his technique of moist-hot air therapy.

Rehabilitation of Professor Lin Kung-Hsiang.

I visit with Professor Lin Kung-Hsiang in China.

I was invited to China in 1982 to lecture at the Central China Agricultural University in Wuhan. During a pre-tour of China and when we were in Guangdong (Canton), I had heard that Professor Lin Kung-Hsiang lived and worked there. I asked if I could see him and my Government guides arranged for Professor Lin to visit me in my hotel room. We spent an evening in intensive conversation as he had very little news of advances in citrus pathology. Professor Lin was a graduate of Cornell University, as was I, and we had a wonderful conversation reminiscing about his days there. He was very frank in telling me of his painful experiences during the Cultural Revolution.

During this visit of April, 1982, he mentioned he was under treatment for cancer. Despite his illness, it was a pleasure for me to witness his strong resolve and optimism. He was hungry for information on all new developments and I briefed him on the work that was currently under way in various areas of citrus pathology, and especially with the greening disease (Huanglongbing - HLB). He was very pleased when I told him that I was indebted to him for his original research on thermotherapy using his moist-hot air treatment where he eliminated the `yellow shoot' (HLB) organism from citrus budwood. I told him that I had used his technique to eliminate the tatter leaf virus from Meyer lemon. I took a picture of him seated in my hotel room and we parted on a note of much mutual affection. That evening, he wrote a letter to me which was delivered to my hotel room the next morning.

Part of his letter was as follows *(His full letter is found on Page 19-20)*:

"I was gratified to know that you adapted my method of heat treatment for eliminating viruses from citrus and your success in propagating nursery trees for commercial citrus growing by using this method had led the California government to pass a law requiring citrus nursery trees be treated with the said method of heat treatment.

I highly appreciate your evaluation on my work on thermotherapy of yellow shoot and blotchy yellows. But for many years, to my disappointment, my work had been neglected partly because of political movements and partly because of the ignorance of the agricultural officials concerned with my work. They usually do not value much of the work done by our own people, and I still have difficulties in pushing ahead my work further.

You will be surprised to know that my method of heat treatment has not yet been adapted on an extension basis. The Chinese Government has great respect for your opinion and whatever suggestions you might make as a foreign specialist.

So, if you should have a chance to mention my work and make your evaluation of the heat treatment method I developed as you did last night, and if you consider it appropriate, it will facilitate my work greatly. We have a Chinese saying "Local ginger does not taste hot."

Figure 4: The picture I took of Prof. Lin

Following my visit with Professor Lin and during my lectures in Wuhan, I presented a comprehensive review on citrus virus diseases to a group of over 100 eminent Chinese Horticulturists and Pathologists (Figure 5). In my lecture on the tatter leaf disease, I cited and emphasized the very important contribution made by Professor Lin in thermotherapy and also much of his other research on the greening (HLB) disease. I also mentioned, and again emphasized, on how valuable his original research was to me for the elimination of the tatter leaf virus from the Meyer lemon by his method of the moist-hot air treatment which permitted the Meyer lemon to be re-introduced back into the Central Valley of California.

We are deeply indebted to Professor Lin Kung-Hsiang for his discovery of the moist-hot air treatment of citrus budwood for the elimination of citrus pathogens.

Figure 5: The group of scientists attending my lecture in Wuhan China in 1982. Professor Chang Wencai shown on bottom left.

Professor Lin died of cancer in June, 1986. I had this photo taken of him during our visit which had sat on my desk for years. In June, 1991 I wrote my friend Prof. Zhao Xue-Yuan if he would like a print of this photo. He replied that he most certainly would appreciate having a picture for his office. He also mentioned that there were a number of people who would enjoy a copy of this picture, one of whom was Dr. Lin's brother; Kung-Hsun Lin. After sending the picture to his brother, I received a letter of thanks containing a number of names who would also enjoy a copy of this picture. He also suggested that Dr. Lin's wife, Lucille Lin, who was then living in Rochester, New York, would surely appreciate that picture. I then had additional copies made and in November, 1991 I sent Mrs. Lin an 8x12 inch picture of her husband. In receiving this picture she wrote me a thank you letter dated December 15, 1991 (see my letter to her on Page 15 and her response on Pages 16-17).

One never knows how our research can evolve into paths that touch our hearts and we can touch one another in ways that are rewarding. Also, how our work in virology, our International Organization of Citrus Virologists on an international scale not only helps others, but can give proper credit to research scientists as we learn from one another.

The rehabilitation of Professor Lin was apparent when in the introductory foreword to the Proceedings of the Asia Pacific International Conference on Citriculture at Chiang Mai, Thailand in February 1990, Dr. Lin was honored by having his picture and the proceedings dedicated, *"In Memory of Professor K. S. Lin"*.

Also, the 13th Proceedings of the International Organization of Citrus Virologists was dedicated to Professor Lin with his picture (below) plus a review of his life and works.

Shown in Figure 6 is a statue of Professor Lin in front of the Citrus Research Institute in Beibei, Chongqing, China.

Dedicated to Lin Kung-Hsiang

(Kung-Hsiang) (1910-1986)

Here is my letter to Mrs. Lucille Lin, the wife of Prof. Lin Kung-hsiang:

UNIVERSITY OF CALIFORNIA, RIVERSIDE

BERKELEY • DAVIS • IRVINE • LOS ANGELES • RIVERSIDE • SAN DIEGO • SAN FRANCISCO  SANTA BARBARA • SANTA CRUZ

Department of Plant Pathology
Riverside, California 92521-0122
Telephone: (909) 787-4117

College of Natural and Agricultural Sciences
Citrus Research Center and Agricultural Experiment Station

November 24, 1991

Mrs. Lucille Lin
84 Lucinda Lane
Rochester, New York 14626

Dear Mrs. Lin,

I recently received a letter from Kung-hsun Lin in reply to my letter (enclosed). He mentioned that you might wish to have a copy of this picture and he gave me your address. I enclose this picture taken while I was visiting with your husband in 1982. I am making more copies perhaps for your sons and daughters in the United States and in Australia. Please let me know how many you may wish to have for them.

As I mentioned in my letter to Hung-hsun Lin, I had the greatest respect for your husband and for his work.

Please let me hear from you - if you have received this at this address and if you would like more prints for your children.

Sincerely yours,

Chester N. Roistacher

Code\letters\lucille.n91

FAX: (909) 787-4294   BITNET: PLPATH@UCRVMS   TELEX: 676427IPM45CA

> *Response to my letter of November 24, 1991 from Lucinda Lin*

Mrs. Lucille Lin
84 Lucinda ln
Rochester, NY 14626

December 15, 1991

Dr. Chester N. Roistacher
Department of Plant Pathology
College of Natural and Agricultural Science
Citrus Research Center and
Agricultural Experiment Station
University of California, Riverside
Riverside, California 92521-0122

Dear Professor Roistacher,

Your letter and the picture of my late husband brought us so much joy and so much memories about him, especially the letter came in the time of the Christmas holiday. We are deeply appreciate your letter and what you did for my husband nine years ago. Perhaps you did not realize how much your visit and your acknowledgment on my husband's work mend to him and to our life. What you did still means a great deal to us to these days.

Because the incompatibilities in political ideal between my husband and the communist regime, my husband was in political trouble in the 50's soon after the communist took over China. In academic area, because his disagreement on the cause of a citrus disease from some Russian expert's theories, he was sent to a farm to do hard labor work for one year, in their words, "to be re-educated". Not only himself was affected by his political believes, the children's future was also diminished in the political environment. All these were seemed to be so unbelievable at the time, the worst was not yet came. During the 60's and 70's, my husband became the target in the worst political struggle in the communist China. For the same reason got my husband in trouble at the first time, he was badly punished both mentally and physically. Young men through mud on his face to show their royalty to the Party. They used ink, wrote political slogans on his back to insult him. Very much like you probably saw on TV back in the 60's and 70's about the Chinese Cultural Revolution, we were living in the most difficult time in our life. Harassments and humiliation became our daily life. In late 70's, my husband was seriously ill in the re-education camp and two of my children, Tony and May, escaped to Hong Kong to seek for a better life. They did not have any future in China, neither school nor job.

That was the political situation and working environment my husband was in for forty years. Needless to say, his work was not be recognized by the government. Being a farmer's son, born and grow up in a citrus farm, watching acres and acres of orange being destroyed by the disease really burn his heart, more so than his own suffering. In 1980, because my husband was no longer be able to do labor work, we were sent back to Canton. He was so sick and almost lost hope. Your 1982 visit changed the matter. Because of your acknowledgement, people, not the officials, started to realize that, may be his work in science should be recognized by our own people. In China, under the communist system, only hand full of decent scientists would openly acknowledge the work done by their colleges who is in trouble. This was the turning point in our life and now you can understand why we appreciate what you did to my husband in 1982.

Today, I am living with my son Tony and his family and all the nightmare is over. Tony graduated from Cornell University in 1979 and now is working for a engineering firm in Rochester as a vice-president. My youngest son Ben-Da is also an engineer who lives in Long Island. My eldest son and two daughters and their families live in Mealboune, Australia. We all enjoy the happiness in this great country, every bit of it. The picture you took nine years ago is probably the best picture in the best time in his life and I put it on the wall in my room so that I feel like seeing him every day. I would like to give each of my five children a copy so they will remember our family's history, a story of their father who fought and suffered for almost forty years. If it is not too much of trouble, please make five copies for me and please kindly let me know the cost of these copies.

Thank you for your kindness and generosity and wish you a merry Christmas and Happy new year.

Lucille Lin

Sincerely,
Lucille Lin

A statue honoring Prof. Lin Kung-Hsiang

Figure 6: This Photo was sent to me by Dr. Chanyong Zhou in February, 2016. The picture was taken in 2013 at a conference at the Citrus Research Institute at Beibei, China. On the left are Professors Changyong Zhou, Zhao Xueyuan and Josy Bové.

Figure 7: This photo was also sent to me by Dr. Chanyong Zhou in January, 2018.

LETTER FROM DR. KUNG-HSIANG LIN - APRIL 10, 1982 Code\letters\lin-482.new

South China Agricultural College
Guangzhow, (Canton) China
April 10, 1982

Dr. C.N. Roistacher
Citrus Research Center
Riverside, Ca.
U.S.A.

Dear Dr. Roistacher,

It was indeed a great pleasure and an excitement to see you and your wife. I didn't know that your wife was with you for otherwise I would have had my wife, Lucille, instead of my son accompany me to see you. Seeing you took me back to the days I very pleasantly spent at Riverside with many of my friends, including the Fawcetts, Klotzs and the Cochrans. I was particularly pleased to know that you are also a Cornellian. Indeed, as you said, we had so much to talk about that we could talk for days or even weeks. I enjoyed greatly the friendship of American people and get along so well with them that they treated me not as an Alien but as an American and an intimate friend. We called each other by first names. At Cornell they called me Frank or just "Lin". At Riverside, L.C. Cochran also called me Frank and Mrs. Frances Hayes called me Franklin. Last night I didn't have the chance to ask you what your fist name is. During the political movements I was severely criticized, one of the reasons being that I was pro-American and anti-Russian. They said that I was more American than Americans, which, of course, was an exaggeration. But anyway, I enjoy making friends with Americans. The political movements haven't changed me a bit in this respect, because I believe what I do is right.

Now let us talk about our work some more. I was gratified to know that you adapted my method of heat treatment for eliminating viruses from citrus and your success in propagating nursery trees for commercial citrus growing by using this method had led the California government to pass a law requiring citrus nursery trees be treated with the said method of heat treatment. I highly appreciate your evaluation on my work on thermotherapy of yellow shoot and blotchy yellows. But for many years, to my disappointment, my work had been neglected partly because of political movements and partly because of the ignorance of the agricultural officials concerned with my work. They usually do not value much of the work done by our own people, and I still have difficulties in pushing ahead my work further. You will be surprised to know that my method of heat treatment has not yet been adapted on an extension basis. The Chinese Government has great respect for your opinion and whatever suggestions you might make as a foreign specialist. So, if you should have a chance to mention my work and make your evaluation of the heat treatment method I developed as you did last night, and if you consider it appropriate, it will facilitate my work greatly. We have a Chinese saying "Local ginger does not taste hot." If is possible and convenient, I shall appreciate greatly having a copy of the recorded tape that you made of our dialogue last night about your work and mine. It would be a most valuable scientific record of our work which I wish my colleagues to learn.

Because of interferences and interruptions by almost incessant political movements, my work on virus-like diseases has been greatly impeded. Aside from yellow shoot and blotchy yellows, we have done very little on other diseases. There is much more to be done or explored. Don't you think it feasible and worth while to have a cooperative project worked out between us, i.e. between Riverside and the Citrus Research Center and South China Agricultural College? At present I have four men working with me and I expect three more to come. And besides, there are three men of the Agricultural Academy of Guangdong Province who are working in close cooperation with me. So you see we have quite a working team. I suppose, such a project if worked out, will be mainly concerned with survey and identification of the disease by indexing with indicator plants. One of my men, Mr.

1

Appendix 2. The full text of Dr. Lin Kung-Hsiang's letter sent to me after our visit in Guangzhow on April 9th 1982 (Following Page).

Shao-jing Chen has a fairly good knowledge of English. He was one of my last students and may take part in such a project (*Mr. Shao-jing Chen attended my course in Wuhan and had just recently been released from the rice fields - he was one of the most knowledgeable participants in my course*). If some arrangement could be made for him to have a short period of training under your guidance at Riverside, I believe it will greatly facilitate the work. But this would not be realized until a fund is made available to finance his trip to and from Riverside and all his expenses during his stay at Riverside, probably for about a period of about three months. The Chinese Government is not in position to do it, I know, for lack of foreign exchange. She is too poor at present.

I hope I might have the opportunity to participate in the Brazil IOCV meeting and besides to revisit Riverside and to tour some other citrus growing areas in the U.S. and elsewhere if possible. I shall appreciate you writing to Dr. Weathers about the matter to see what can be done.

I am sending you a little iron-cut flower as a souvenir in care of Mr. Chen, who is going to attend your lectures in Hangkow (Wuhan). I am also asking him to bring an album for you to autograph for me to keep as a memory. Practically all the American plant pathologists have autographed it during my tour in the U.S. in 1941 and afterwards. You may autograph on page 48 which is still blank. It may be interesting for you to see the autographs of those people including Dr. Webber, Prof. Whetzel, Drs. L.R. Jones of Wisconsin and Stakman of Minnesota. The Riverside people autographed on pp. 93-111, 42 and 157. The album may be returned to me also through Mr. Chen when he comes back after attending your lectures.

With best wishes for your successful trip and visit in China. Best regards to Mrs. Roistacher, Dr. Weathers Klotz and other friends.

Sincerely yours,

Frank

Kung-Hsiang Lin

I am also sending a Chinese Calligraphic work, a poem written by myself which you will please give to L. C. Cochran when you see him, as a memory of our friendship. Thanks.

Shall appreciate receiving books or papers on Rickettsia, Moliente & Greeninicuta.

Appendix 1. The original paper on using moist-hot air for eliminating a pathogen from citrus.

A PRELIMINARY STUDY ON THERMOTHERAPY OF YELLOW SHOOT DISEASE OF CITRUS

Lin Kung-hsiang, Lo Hsueh-hai

(South China Agricultural College)

Yellow shoot affected nursery trees of Ponkan (*Citrus ponkan* Tanaka) were dug up and treated with water vapor saturated hot air in an electric heating chamber at 48°, 49° and 50°C for 31, 35 and 40 minutes. Most of the trees replanted after the treatment completely recovered and remained healthy even after 28 months. Three year old cheokan (*C. tankan* Tanaka) trees affected with the same disease were similarly treated at 48, 49 and 50°C for 45, 50, 55, 60 and 65 minutes and at 51°C for 35, 40, 45, 50 and 55 minutes. All the treated trees completely recovered after replanting. Heat injury was apparent at higher temperatures.

The results obtained show that it is possible to free citrus nursery trees from infection by yellow shoot disease by treating them with water vapor saturated hot air at suitable temperatures for appropriate lengths of time.

华 南 农 业 大 学
SOUTH CHINA AGRICULTURAL UNIVERSITY
GUANGZHOU CHINA

OBITUARY NOTICE

Prof. Lin Kong-Xiang passed away peacefully on June 6, 1985 in Guangzhou, China. A Memorial Service will be held at Guangzhou's Funeral Home on June 11, 1985.

— Funeral Committee of Prof. Lin Kong-Xiang

Books to be Published by Dr. C. N. Roistacher

GROWING UP IN NEW YORK

From birth to U.S. Navy (1924 – 1943 - to be published)

My aunts and uncles in Poland. All but mother and two Aunt swere killed in the Holocaust. My Mother and Father When young.

Wedding picture   *Baby Chet and Sister Miriam*   *Miriam, Chet and Harry*   *Harry, Chet and Scotty in Bronx*

Chet in Public school   *Brother Harry was run over and dies*   *My Joe DiMaggio story*   *I leave for the Navy – Nov. 1942*

My Life in Agriculture Book One 1943 - 1945

The story of how this collection of seeds permitted Chet to enter Cornell University after an initial rejection. This was another miracle in his life.

Chet retired from the Navy in 1943. In 1944 he left N.Y. City and engineering and went to Farm School in Bucks County, PA.

1945-1949: Chet called this period of four years at Cornell University "My Golden Years".

During the summer of 1945 Chet hitched across country to Los Angeles. He worked in the UCLA orchards, and he falls in love.

My Life in Agriculture Book One 1943 - 1945 FROM FARM SCHOOL TO CORNELL
MY GOLDEN YEARS

During the summer of 1946 Chet worked on the Fleming farm in upstate NY, a requirement for practical farm work.

A colorful grape leaf determines Chet's future and career in Plant Pathology. Dr. Baker offers him a full time job at UCLA after he graduates Cornell

In 1948 he met Dr. Kenneth Baker at Cornell and worked for him at UCLA that summer. He built a soil steamer.

Chet met Roberta in 1950. They fall in love and are married.   Shown in their apartment with their kitty "Cisco".

My GOLDEN YEARS - 1951 SMITHFIELD FARM
IN 1951 CHET FARMS IN DOYLESTOWN PENNSYLVANIA.

Smithfield Farm: Showing our beautiful home in the summer and fall.

Winter scenes: Showing the road into our farm after a snowfall.

More winter scenes at Smithfield farm.

BOOK 1   1951 – 1953 FROM SMITHFIELD FARM
TO THE UNIVERSITY OF CALIFORNIA

1951-1952 – Smithfield farm. We work and manage this farm for Dr. Vorhous. Baby Robin is born on this farm

1953 We return to California and buy our house on Carver Court.

Chet begins working on citrus decay problems with Dr. Leo Klotz

MY LIFE IN AGRICULTURE
BOOK 2
1953 -1969

1965 - A love story: Jean and I are maried *1965 - Our new home* *1967- We buy a mountain cabin*

Drs. Calavan and Ruether-Outstanding leaders *Our UC mix wins many awards* *1961- planting the first Foundation block tree*

1966 –Chet and Jean in Paris *1969 - Poet's day in Wakayama, Japan* *1969 - IOCV Conference. The Hiroshima memorial*

MY LIFE IN AGRICULTURE
BOOK 3 - THE 1970'S

Chapter 1. The crazy viroids. They are so small yet they are very destructive

Chapter 2. Thermotherapy - All Meyer lemons are my Children.

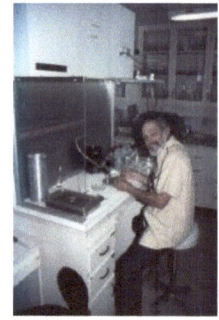  

Chapter 3. Shoot tip grafting, the magic cut that elimated citrus pathogens.   Chapter 4. Venzueala. I tried to warn them.

Chapter 5. South Africa – The power of Folk music.   Chapter 6. Family pictures in the 1970's

A BRIEF BIOGRAPHY OF CHESTER N. ROISTACHER
(CURRICULUM VITAE)
1951 to 2012

Cornell University: Graduated Cornell University 1949 with a major in Plant Pathology.

University of California: Responsible for citrus virus detection and eradication in the California Citrus Clonal Protection Program, University of California at Riverside for 26 years (1960-1986). Retired in 1986

Consultant: For over 50 years had been a consultant and lecturer for the World Bank, the Food and Agriculture Organization (FAO) of the United Nations, US-AID and GTZ. I have consulted in Spain *(1975, 1991)*, South Africa *(1977, 1985, 1993, 1999)*; Venezuela *(1979)*; Mexico *(1976, 1979, 1980, 1992, 1997, 2004, 2007)*; China *(1982, 1986, 1993, 1995, 2008)*; Taiwan *(1985)*; Florida (1987,1997); Republic of the Maldives (1987-1989); Turkey (1989, 1990, 1991, 1997); Uruguay and Argentina (1990, 1991, 1998, 2005); Belize *(1991, 1995)*; Costa Rica (1991); Thailand (1994, 1995); Nepal (1996); Trinidad (1996); Tropicana-Florida (1996, 1997); Bermuda (1997); Brazil *(1997, 1998, 2001)*; Cuba *(1998)*; Jamaica *(1999)*; Puerto Rico *(2000);* Egypt *(2001, 2002, 2003, 2005, 2006)*; Peru *(1987, 2003, 2007);* Suriname *(2001)*; Portugal *(2005);*Cyprus *(2001)*, Morocco *(2010)*. In connection with citrus disease research I have also visited: India, Australia, Oman, Israel, Crete, France, Japan, Sicily and the Philippine Islands. Reports of over 15 of these consultancies can be seen in www.ecoport.org as eArticles.

Teacher and lecturer: At the Institut Agronomique Mèditerranèen at Valenzano, Bari, Italy for 26 consecutive years (1986-2011).

Chairman of the International Organization of Citrus Virologists (IOCV). 1989-1992. Secretary of IOCV 1995-2007

1991 Published the "Handbook for Detection and Diagnosis of Graft Transmissible Pathogens of Citrus". This was a joint effort of the FAO and the IOCV. The book of 286 pages with 250 color illustrations is a standard reference for diagnosis of citrus virus and virus-like diseases.

1995 Published: "A Historical Review of the Major Graft-transmissible Diseases of Citrus" (89 Pgs. with color photographs.

Publications: Has authored and co-authored over 200 publications since 1951. The major subjects were on citrus virus and virus-like diseases and their control. Pioneered in thermotherapy for citrus, and with Dr. Luis Navarro we developed the shoot tip grafting method for elimination of citrus viruses. Developed rapid indicators and superior plants for the detection of various citrus pathogens (exocortis, cachexia, psorosis, Dweet mottle virus etc.). Developed Sodium hypochlorite for the disinfection of tools. Developed virus-free Meyer lemon. Predicted the viroid nature of the cachexia disease. Did extensive vector transmission work with tristeza and *Aphis gossypii* and warned on the threat of tristeza and *Toxoptera citricida* to Central America, Mexico, the United States and Venezuela. Developed cross protection methods against the citrus tristeza virus by passage of the virus through *Passiflora* species. This is being used successfully for cross protection in Peru.

Internet slide shows: Developed over 50 slide shows and 15 e Articles on citrus virus and virus like diseases for the United Nations website: www.ecoport.org (Click on Photos/slide shows).

Honorary Degree: On Sept. 10, 1999 was presented with an Honorary Doctorate Degree (DSc) on the recommendation of the Council of the Senate of the University of Pretoria, South Africa.

Awarded as Fellow of the International Organization of Citrus Virologists in November, 2004 at the 16th Conference of the IOCV in Monterey Mexico along with Drs. Steve Garnsey and Josy Bove. This was the first occasion for the awarding of a Fellow for IOCV.

References cited.

Fawcett, H.S. and L.C. Cochran. 1941.
    Resistance of citrus tissue and psorosis virus A to heat. Phytopathology 31:861 (Abstr.).

Price W. C. and L.C. Knorr. 1956.
    Kinetics and thermal destruction of citrus tissues in relation to the virus disease problem. Phytopathology 46:657-661

Grant, T.J. 1957.
    Effect of heat treatments on tristeza and psorosis viruses in citrus. Plant Dis. Reptr. 41:232-234.

Grant, T.J., J.W. Jones and G. G. Norman. 1960.
    Present status of heat treatment of citrus viruses. Proc Fla. State Hort. Soc. 72:45-48.

Roistacher, C.N., E.C. Calavan, E.M. Nauer and W. Reuther. 1972.
    Virus free Meyer lemon trees. Citrograph 57: 250, 270-271.

Roistacher, C.N., and E.C. Calavan. 1972.
    Heat tolerance of preconditioned citrus budwood for virus inactivation. p. 256-261. *In* Proc. 5th Conf. IOCV, Univ. Fla. Press, Gainesville.

All Meyer Lemons are my Children

www.ingramcontent.com/pod-product-compliance
Lightning Source LLC
Chambersburg PA
CBHW040415220526
45473CB00004B/1246